D0776768

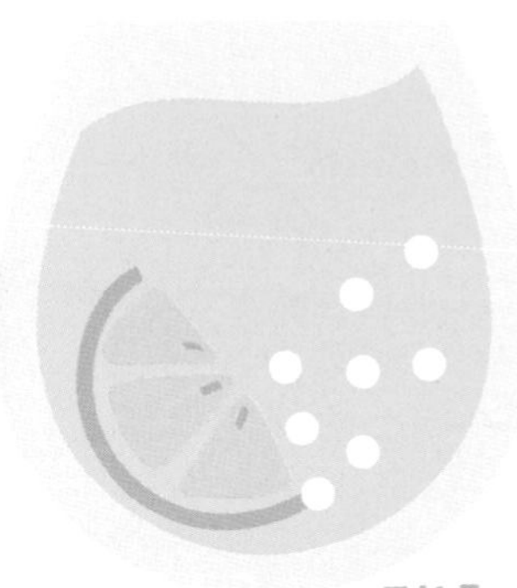

THE LITTLE BOOK OF

COCKTAILS

First published in Great Britain in 2022 by Pyramid,
an imprint of Octopus Publishing Group Ltd
Carmelite House, 50 Victoria Embankment, London EC4Y 0DZ
www.octopusbooks.co.uk

Distributed in the US by
Hachette Book Group
1290 Avenue of the Americas
4th and 5th Floors
New York, NY 10104

Distributed in Canada by
Canadian Manda Group
664 Annette St.
Toronto, Ontario, Canada M6S 2C8

ISBN 978-0-7537-3510-7

A CIP catalogue record for this book is available from the British Library

Printed and bound in China

10 9 8 7 6 5 4 3 2 1

Publisher: Lucy Pessell
Designer: Hannah Coughlin
Editor: Sarah Kennedy
Editorial Assistant: Emily Martin
Production Controller: Serena Savini

The measure that has been used in the recipes is based on a bar jigger, which is 25 ml (1 fl oz). If preferred, a different volume can be used, providing the proportions are kept constant within a drink and suitable adjustments are made to spoon measurements, where they occur.

Standard level spoon measurements are used in all recipes.
1 tablespoon = one 15 ml spoon
1 teaspoon = one 5 ml spoon

This book contains cocktails made with raw or lightly cooked eggs. It is prudent for more vulnerable people to avoid uncooked or lightly cooked cocktails made with eggs.

TO GIN OR NOT TO GIN? SILLY QUESTION

The G&T is the elixir of life to gin fans but, what with there being literally thousands of gins out there to try, we've run out of excuses not to experiment a bit and let gin work its magic in our cocktails.

With Gatsby-esque classics like the Gin Rickey, modern classics like the Bramble and gincredibly ginspiring twists like the Elderflower Mojito and Espresso Gin-tini, this book has a deliciously drinkable cocktail for everyone for every occasion.

So, whether you're looking for something sophisticated straight out of a speakeasy, a Monday martini made in heaven or jug of something ginny for a garden get-together, *The Little Book of Gin Cocktails* has it all.

MARTINIS

BREAKFAST MARTINI

1¾ measures gin

½ measure Cointreau

¾ measure lemon juice

1 tsp orange marmalade

orange and a small slice of toast, to garnish

Add all the ingredients to a cocktail shaker and give the liquid a quick stir to break up the marmalade.

Shake vigorously with cubed ice and double strain into a chilled martini glass.

Garnish with an orange twist and a small slice of toast.

SOUTHSIDE

2 measures gin

4 teaspoons lime juice

4 teaspoons sugar syrup

5 mint leaves plus extra, to garnish

Add all the ingredients to a cocktail shaker with some ice cubes.

Shake and strain into a martini glass.

Garnish with a mint leaf and serve.

GIBSON MARTINI

2½ measures gin

½ measure dry vermouth

cocktail onions, to garnish

Add the gin and dry vermouth to a cocktail shaker, and fill with cubed ice.

Stir for 30 seconds, and strain into a chilled martini glass.

Garnish generously with cocktail onions.

DOOBS
MARTINI

1¾ measures gin

1 measure sloe gin

¾ measure dry vermouth

4 dashes orange bitters

orange, to garnish

Add all the ingredients to a cocktail shaker and fill with cubed ice.

Stir for 30 seconds, and strain into a chilled martini glass.

Garnish with an orange twist.

CLASSIC MARTINI

makes 2

1 measure dry vermouth

6 measures gin

stuffed green olives, to garnish

Put 10–12 ice cubes into a mixing glass.

Pour over the vermouth and gin, then stir (never shake) vigorously and evenly, without splashing.

Strain into 2 chilled martini glasses, garnish each with a green olive and serve.

HANKY PANKY

2 measures gin

1 measure sweet vermouth

1 tablespoon Fernet Branca

orange, to garnish

Add all the ingredients to a cocktail shaker and fill with cubed ice.

Stir for 30 seconds, and strain into a chilled martini glass.

Garnish with an orange twist.

VESPER MARTINI

2½ measures gin

1 measure vodka

½ measure Lillet Blanc

lemon, to garnish

Add all the ingredients to a cocktail shaker and fill the top half of it with ice.

Shake vigorously and double strain into a chilled martini glass.

Garnish with a lemon twist.

GIN GARDEN MARTINI

makes 2

4 measures gin

2 measures pressed apple juice

1 measure elderflower cordial

½ cucumber, peeled and chopped, plus extra slices, to garnish

Muddle the cucumber in the bottom of a cocktail shaker with the elderflower cordial.

Add the gin, apple juice and some ice cubes.

Shake and double-strain into 2 chilled martini glasses, garnish with peeled cucumber slices and serve.

ESPRESSO GIN-TINI

1½ measures gin

1 measure coffee liqueur

1 measure fresh espresso coffee

½ measure sugar syrup

3 coffee beans, to garnish

Add all the ingredients to a cocktail shaker.

Shake vigorously and double strain into a chilled martini glass.

Garnish with the 3 coffee beans.

LYCHEE MARTINI

1½ measures gin

1 measure lychee liqueur

½ measure lychee syrup (from the tin)

¾ measure lemon juice

lychees (tinned), to garnish

Add all the ingredients to a cocktail shaker, shake vigorously and double strain into a chilled martini glass.

Garnish with lychees.

RED RUM

makes 2

1 measure sloe gin

handful of redcurrants plus extra, to garnish

4 measures Bacardi 8-year-old rum

1 measure lemon juice

1 measure vanilla syrup

Muddle the sloe gin and redcurrants together in a cocktail shaker.

Add the rum, lemon juice, vanilla syrup and some ice cubes.

Shake and double-strain into 2 chilled martini glasses, garnish with redcurrants and serve.

FLUTES & COUPETTES

ARMY & NAVY

2 measures gin

1 measure lemon juice

½ measure orgeat syrup

lemon, to garnish

Add all the ingredients to a cocktail shaker.

Shake vigorously with cubed ice and double strain into a chilled coupette glass.

Garnish with a lemon twist.

BEES KNEES

2 measures gin

1 measure lemon juice

½ measure honey

lemon, to garnish

Add all the ingredients to a cocktail shaker.

Shake vigorously with cubed ice, double strain into a chilled coupette glass and garnish with a lemon twist.

CLOVER CLUB

2 measures gin

¾ measure lemon juice

¾ measure sugar syrup

5 raspberries

½ measure egg white

Add all the ingredients to a cocktail shaker and vigorously dry-shake without ice for 10 seconds.

Take the shaker apart, add cubed ice and shake again vigorously.

Strain into a coupette glass and garnish with raspberries.

FRENCH 75

makes 2

2 measures gin
6 teaspoons lemon juice
6 teaspoons sugar syrup
8 measures Champagne
lemon twist, to garnish

Add the gin, lemon juice and sugar syrup into a cocktail shaker and shake.

Strain into 2 flute glasses and top with the chilled Champagne.

Garnish with a lemon twist and serve.

GIMLET

2½ measures gin

½ measure lime cordial

½ measure lime juice

lime, to garnish

Add all the ingredients to a cocktail shaker, shake vigorously and strain into a chilled coupette glass.

Garnish with a lime twist.

PERFECT LADY

1½ measures gin

¾ measure lemon juice

½ measure peach liqueur

½ measure egg white

Add all the ingredients to a cocktail shaker and vigorously dry-shake without ice for 10 seconds.

Take the shaker apart, add cubed ice and shake vigorously.

Double strain into a chilled coupette glass.

No garnish.

FRENCH PINK LADY

2 measures gin

1 measure triple sec

4 raspberries

3 teaspoons lime juice

1 teaspoon pastis

lime wedge, to garnish

Add all the ingredients except for the lime wedge to a cocktail shaker and muddle.

Fill the shaker with ice cubes and shake, then strain into a coupette glass.

Garnish with a lime wedge and serve.

WHITE LADY

1½ measures gin

1 measure Cointreau

¾ measure lemon juice

lemon, to garnish

Add all the ingredients to a cocktail shaker, shake vigorously and double strain into a chilled coupette glass.

Garnish with a lemon twist.

SATAN'S WHISKERS

1½ measures gin

½ measure orange curaçao

½ measure sweet vermouth

½ measure dry vermouth

1½ measures orange juice

2 dashes orange bitters

Add all the ingredients to a cocktail shaker, shake vigorously and double strain into a chilled coupette glass.

No garnish.

MARTINEZ

2 measures gin

1 measure sweet vermouth

1 teaspoon Maraschino

2 dashes Angostura bitters

orange twist, to garnish

Add all the ingredients to a cocktail shaker and fill with cubed ice.

Stir for 30 seconds and strain into a chilled coupette glass.

Garnish with an orange twist.

CORPSE REVIVER NO. 2

1 measure gin

1 measure lemon juice

1 measure Lillet Blanc

1 measure Cointreau

2 drops absinthe

lemon, to garnish

Add all the ingredients to a cocktail shaker.

Shake vigorously with cubed ice and double strain into a chilled coupette glass.

Garnish with a lemon twist.

LONDON CALLING

1½ measures gin

¾ measure lemon juice

¾ measure Fino sherry

½ measure sugar syrup

2 dashes Angostura bitters

lemon, to garnish

Add all the ingredients to a cocktail shaker, shake vigorously and strain into a chilled coupette glass.

Garnish with a lemon twist.

ROYAL COBBLER

3 teaspoons gin

3 teaspoons Fino sherry

3 teaspoons raspberry and pineapple syrup

2 teaspoons lemon juice

3 measures Prosecco

raspberry, to garnish

Add the gin, Fino sherry, raspberry and pineapple syrup and lemon juice to a cocktail shaker.

Shake and strain into a flute glass and top with chilled Prosecco.

Garnish with a raspberry or orange twist and serve.

GOLDEN DAWN

1 measure gin

1 measure Calvados

1 measure apricot brandy

1½ measures orange juice

2 teaspoons grenadine

2 dashes Angostura bitters

lemon, to garnish

Add all the ingredients except the grenadine to a cocktail shaker, shake vigorously and strain into a coupette glass.

Carefully add the grenadine, allowing it to float to the bottom of the glass to create a "sunrise" effect.

Garnish with a lemon twist.

LADY OF LEISURE

1 measure gin

½ measure Chambord

½ measure Cointreau

¼ measure lemon juice

1 measure pineapple juice

orange, to garnish

Add all the ingredients to a cocktail shaker, shake vigorously and strain into a coupette glass.

Garnish with an orange twist.

ROCKS & WINE GLASSES

BRAMBLE

2 measures gin

1 measure lemon juice

½ measure sugar syrup

½ measure crème de mûre

lemon and blackberries, to garnish

Fill a rocks glass with crushed ice, packing it in tightly.

Add the gin, lemon juice and sugar syrup and stir briefly.

Slowly drizzle over the crème de mûre, so that it creates a "bleeding" effect down through the drink.

Top with more crushed ice and garnish with blackberries and a lemon wedge.

NEGRONI

1 measure gin

1 measure sweet vermouth

1 measure Campari

orange wedges, to garnish

Add all the ingredients to a rocks glass filled with ice cubes and stir.

Garnish with an orange twist and serve.

LYCHEE NEGRONI

¾ measure gin

1½ measures bianco vermouth

½ measure Campari

1 measure lychee juice

cucumber, to garnish

Add all the ingredients to a rocks glass full of cubed ice, stir briefly and garnish with a slice of cucumber.

GIN RICKEY

2 measures gin

¾ measure lime juice

½ measure sugar syrup

soda water, to top

lime, to garnish

Add all the ingredients to a large wine glass full of cubed ice, stir briefly and garnish with a lemon wedge and a sprig of mint.

GINNY GIN FIZZ

2 measures gin
1 camomile tea bag
1 measure sugar syrup
1 measure lemon juice
3 teaspoons egg white
3 measures soda water
lemon twist, to garnish

Place the gin and camomile tea bag in a cocktail shaker and leave to infuse for 2 minutes.

Remove the tea bag, add the sugar syrup, lemon juice and egg white.

Fill the shaker with ice cubes.

Shake and strain into a wine glass filled with ice cubes and top with the soda water.

Garnish with a lemon twist and serve.

GIN CUP

2 measures gin

¾ measure lemon juice

½ measure sugar syrup

3 mint sprigs, plus extra to garnish

Muddle the mint and sugar syrup in a rocks glass.

Fill the glass with crushed ice, add the gin and lemon and churn vigorously until a frost begins to form on the glass.

Garnish with mint sprigs.

SLOE GIN SOUR

2 measures sloe gin

1 measure lemon juice

1 measure sugar syrup

½ measure egg white

2 dashes Peychaud's bitters

orange, to garnish

Add all the ingredients to a cocktail shaker and dry-shake without ice for 10 seconds, take the shaker apart and add cubed ice.

Shake vigorously and double strain into an old fashioned glass filled with cubed ice.

Garnish with an orange slice.

ORANGE BLOSSOM

2 measures gin

2 measures pink grapefruit juice

2 teaspoons orgeat

2 dashes Angostura bitters

4 orange slices

orange, to garnish

Muddle the orange slices and orgeat in a rocks glass, add the remaining ingredients, fill with crushed ice and churn.

Top with more crushed ice and garnish with orange wedges.

P&T

1½ measures pink gin

½ measure strawberry liqueur

½ measure lemon juice

equal parts tonic and soda water, to top

rosemary and grapefruit, to garnish

Add all the ingredients to a large wine glass full of cubed ice, stir briefly and garnish with a sprig of rosemary and a slice of grapefruit.

STRAWBERRY FIELDS

2 measures gin

1 camomile tea bag

1 measure strawberry purée

2 teaspoons lemon juice

1 measure double cream

soda water, to top

strawberry, to garnish

Add the gin and tea bag to a cocktail shaker and leave to infuse for 2 minutes, stirring occasionally.

Remove the tea bag and add the rest of the ingredients to the shaker.

Shake vigorously, strain into a wine glass full of cubed ice, top with soda water and garnish with a strawberry.

GIN CUCUMBER COOLER

2 measures gin

5 mint leaves

5 slices cucumber

3 measures apple juice

3 measures soda water

mint sprig, to garnish

Add the gin, mint and cucumber to a rocks glass and gently muddle.

Leave to stand for a couple of minutes, then add the apple juice, soda water and some ice cubes.

Garnish with a mint sprig.

HIGHBALLS

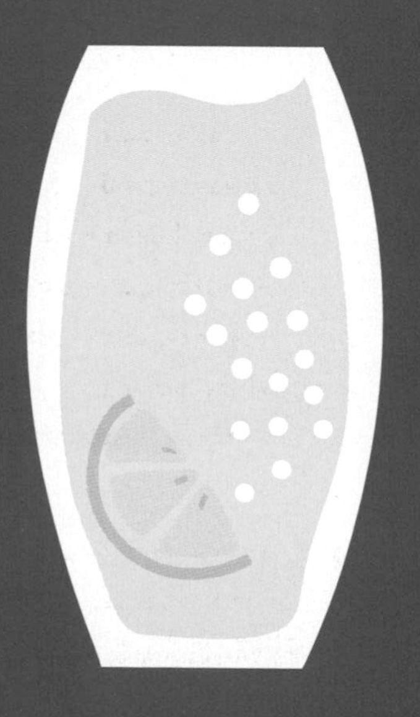

SAKE COLLINS

1 measure gin

2 measures sake

½ measure lemon juice

¾ measure sugar syrup

½ measure grapefruit juice

soda water, to top

cucumber and grapefruit, to garnish

Add all the ingredients except the soda water to a cocktail shaker.

Shake vigorously, strain into a highball glass full of cubed ice and top with soda.

Garnish with a slice of grapefruit and a cucumber ribbon.

TOM COLLINS

2 measures gin

1 measure sugar syrup

1 measure lemon juice

4 measures soda water

lemon wedge, to garnish

Put the gin, sugar syrup and lemon juice into a cocktail shaker and fill with ice cubes.

Shake and strain into a highball glass full of ice cubes and top with the soda water.

Garnish with a lemon wedge.

HEDGEROW COLLINS

1½ measures gin

½ measure crème de mûre

1 teaspoon Campari

¾ measure lemon juice

½ measure sugar syrup

soda water, to top

lemon and blackberry, to garnish

Add all the ingredients except the soda water to a highball glass filled with cubed ice.

Stir gently, top with soda and garnish with a lemon wedge and a blackberry.

BERRY COLLINS

makes 2

8 raspberries plus extra, to garnish

8 blueberries plus extra, to garnish

1–2 dashes strawberry syrup

4 measures gin

4 teaspoons lemon juice

sugar syrup, to taste

soda water, to top up

lemon slices, to garnish

Muddle the berries and strawberry syrup in the bottom of two highball glasses, then fill each glass with crushed ice.

Add the gin, lemon juice and sugar syrup.

Stir, then top with the soda water.

Garnish with berries and lemon slices and serve.

CAMOMILE COLLINS

2 measures gin

1 camomile tea bag

1 measure lemon juice

1 measure sugar syrup

4 measures soda water

lemon slice, to garnish

Pour the gin into a highball glass and add the tea bag.

Stir the tea bag and gin together, for about 5 minutes, until the gin is infused with the camomile flavour.

Remove the tea bag and fill the glass with ice cubes.

Add the remaining ingredients and garnish with a lemon slice.

BRITISH MOJITO

2 measures gin

¾ measure lime juice

½ measure elderflower cordial

6–8 mint leaves

soda, to top

lime and mint, to garnish

Add all the ingredients except the soda water to a highball glass.

Fill the glass with crushed ice, and churn with a bar spoon.

Add a splash of soda water, and top with more crushed ice.

Garnish with a lime wedge and a mint sprig.

ELDERFLOWER MOJITO

1½ measures gin

1 measure elderflower liqueur

½ measure lemon juice

equal parts tonic and soda water, to top

cucumber and mint, to garnish

Add all the ingredients to a highball full of crushed ice and churn.

Top with more crushed ice and tonic as needed and garnish with a cucumber slice and a sprig of mint.

PIMM'S COCKTAIL

makes 2

2 measures Pimm's No. 1 Cup

2 measures gin

4 measures lemonade

4 measures ginger ale

to garnish:

cucumber strips

blueberries

orange slices

Fill 2 highball glasses with ice cubes.

Add all the ingredients, one by one in order, over the ice.

Garnish with cucumber strips, blueberries and orange slices and serve.

SINGAPORE SLING

1 measure gin

1 measure Cointreau

½ measure Bénédictine

½ measure cherry brandy

¾ measure lemon juice

soda water, to top

lemon and cocktail cherry, to garnish

Add all the ingredients except the soda water to a cocktail shaker.

Shake vigorously, strain into a hurricane glass filled with cubed ice and top with soda water.

Garnish with a lemon wedge and cocktail cherry.

GIN & IT

1½ measures gin

1½ measures sweet vermouth

orange, to garnish

Add all the ingredients to a rocks glass filled with cubed ice, stir briefly and garnish with a slice of orange.

MANGO RICKY

5 basil leaves plus extra, to garnish

2 lime wedges

1 measure mango-infused gin

2 teaspoons sugar syrup

2 measures soda water

Roughly tear the basil leaves and add to a highball glass.

Squeeze the lime wedges into the glass and then add them to the glass.

Add the gin, sugar syrup and soda water, then top the glass with crushed ice.

Garnish with basil leaves and serve.

FIX

2 measures gin

1 measure lemon juice

¾ measure sugar syrup

seasonal fruit, to garnish

Add all the ingredients to a rocks glass filled with crushed ice.

Churn, and garnish with seasonal fruit of your choosing.

G & TEA

1½ measures gin

¾ measure peach liqueur

¾ measure lemon juice

½ measure sugar syrup

2 measures cold breakfast tea

lemon and rosemary, to garnish

Add all the ingredients to a cocktail shaker, shake vigorously and strain into a chilled highball glass filled with cubed ice.

Garnish with a lemon wedge and a sprig of rosemary.

CHERRY JULEP

¾ measure gin

¾ measure sloe gin

¾ measure cherry brandy

¾ measure lemon juice

1 teaspoon sugar syrup

1 teaspoon grenadine

lemon and mint, to garnish

Add all the ingredients to a highball glass filled with crushed ice, and churn vigorously.

Top with more crushed ice and garnish with a wedge of lemon and a mint sprig.

LONG ISLAND ICED TEA

makes 2

1 measure vodka
1 measure gin
1 measure white rum
1 measure tequila
1 measure Cointreau
1 measure lemon juice
cola, to top up
lemon slices, to garnish

Put the vodka, gin, rum, tequila, Cointreau and lemon juice in a cocktail shaker with some ice cubes and shake to mix.

Strain into 2 highball glasses filled with ice cubes and top with cola.

Garnish with lemon slices.

JUGS & BOWLS

MULLED GIN

6 measures apple juice

2 measures gin

1 cinnamon stick

2 star anise

4 whole cloves

1 dash lime juice

Add all the ingredients to a saucepan.

Heat the liquid gently (don't let it boil) to steep the spices.

After 10 minutes serve in heatproof glasses or mugs.

TWISTED SANGRIA

4 measures gin

6 measures apple juice

2 measures lemon juice

2 measures elderflower cordial

6 measures white wine

6 measures soda water

apple, lemon and mint, to garnish

Add all the ingredients to a jug filled with ice cubes and stir.

Garnish with apple and lemon slices and mint leaves.

LANGRA AND TONIC

makes 1 large jug

200 ml (7 fl oz) gin
4 measures mango juice
2 measures lemon juice
2 measures sugar
200 ml (7 fl oz) tonic water
lemon wheels, to garnish

Add all the ingredients to a jug filled with ice cubes and stir.

Garnish with lemon wheels and serve.

GARDEN COOLER

makes 1 large punch bowl

700 ml (23½ fl oz) London dry gin

500 ml (17 fl oz) lemon juice

250 ml (8 fl oz) sugar syrup

250 ml (8 fl oz) elderflower cordial

500 ml (17 fl oz) apple juice

500 ml (17 fl oz) green tea, cooled

500 ml (17 fl oz) mint tea, cooled

500 ml (17 fl oz) soda water

peach slices, to garnish

Add all the ingredients to a punch bowl filled with ice cubes and stir.

Garnish with peach slices and serve.

EARL'S PUNCH

makes 1 large jug

4 measures gin

6 measures Earl Grey tea, chilled

6 measures pink grapefruit juice

6 measures soda water

1 measure sugar syrup

to garnish:

pink grapefruit slices

black cherries

Add all the ingredients to a jug filled with ice cubes and stir.

Garnish with pink grapefruit slices and black cherries and serve.

BAR BASICS & TECHNIQUES

THE BASICS

Good cocktails, like good food, are based around quality ingredients. As with cooking, using fresh and homemade ingredients can often make the huge difference between a good drink and an outstanding drink. All of this can be found in department stores, online or in kitchen shops.

Ice

This is a key part of cocktails and you'll need lots of it. Purchase it from your supermarket, or freeze big tubs of water then crack them up to use in your drinks. If you're hosting a big party and want to serve some punches, which will need lots of ice, it may be worthwhile to find out if you have a local ice supplier that supplies catering companies, as this can be much more cost-effective.

Citrus juice

It's important to use fresh citrus juice; bottled versions taste awful and will not produce good drinks.
Store your fruit at room temperature. Look for a soft-skinned fruit for juicing, which you can do with a juicer or citrus press. You can keep fresh citrus juice for a couple of days in the refrigerator, sealed to prevent oxidation.

Sugar syrup

You can buy sugar syrup or you can make your own. The most basic form of sugar syrup is made by mixing caster sugar and hot water together, and stirring until the sugar has dissolved. The key is to use a 1:1 ratio of sugar to liquid. White sugar acts as a flavour enhancer, while dark sugars have unique, more toffee-like flavours that work well with dark spirits.

BASIC SUGAR SYRUP RECIPE

Makes 1 litre (1¾ pints)

1 kg (2 lb) caster sugar

1 litre (1¾ pints) hot water

Dissolve the caster sugar in the hot water.

Allow to cool.

The sugar syrup will keep in a sterilized bottle stored in the refrigerator for up to two weeks.

CHOOSING GLASSWARE

There are many different cocktails, but they all fall into one of three categories: long, short or shot. Long drinks generally have more mixer than alcohol, often served with ice and a straw. The terms "straight up" and "on the rocks" are synonymous with the short drink, which tends to be more about the spirit, often combined with a single mixer at most. Finally, there is the shot which is made up mainly from spirits and liqueurs, designed to give a quick hit of alcohol. Glasses are tailored to the type of drinks they will contain, each of which is described opposite.

Champagne flute

Used for Champagne or Champagne cocktails, the narrow mouth of the flute helps the drink to stay fizzy.

Champagne saucer

A classic glass, but not very practical for serving Champagne as the drink quickly loses its fizz.

Margarita or Coupette glass

When used for a Margarita, the rim is dipped in salt. Also used for daiquiris and other fruit-based cocktails.

Highball glass

Suitable for any long cocktail, such as a Long Island Iced Tea.

Collins glass

This is similar to a highball glass but is slightly narrower.

Wine glass

Sangria is often served in one, but they are not usually used for cocktails.

Old-Fashioned glass

Also known as a rocks glass, this is great for any drink that's served on the rocks or straight up.

Shot glass

Often found in two sizes – for a single or double measure. They are ideal for a single mouthful.

Balloon glass

Often used for fine spirits. The glass can be warmed to encourage the release of the drink's aroma.

Hurricane glass

Mostly found in beach bars, used for creamy, rum-based drinks.

Boston glass

Often used by bartenders for mixing fruity drinks.

Toddy glass

A toddy glass is generally used for a hot drink, such as Irish Coffee.

Sling glass

This has a very short stemmed base and is most famously used for a Singapore Sling.

Martini glass

Also known as a cocktail glass, its thin-neck design makes sure your hand can't warm the glass or the cocktail.

USEFUL EQUIPMENT

Some pieces of equipment, such as shakers and the correct glasses, are vital for any cocktail enthusiast. Below is a wish list of things to have to hand for anyone who wants to regularly make decent cocktails.

Shaker

The Boston shaker is the most simple option, but it needs to be used in conjunction with a hawthorne strainer. Alternatively you could choose a shaker with a built-in strainer.

Measure or jigger

Single and double measures are available and are essential when you are mixing ingredients so that the proportions are always the same. One measure is 25 ml or 1 fl oz.

Mixing glass

A mixing glass is used for those drinks that require only a gentle stirring before they are poured or strained.

Hawthorne strainer

This type of strainer is often used in conjunction with a Boston shaker, but a simple tea strainer will also work well.

Bar spoon

Similar to a teaspoon but with a long handle, a bar spoon is used for stirring, layering and muddling drinks.

Muddling stick

Similar to a pestle, which will work just as well, a muddling stick, or muddler, is used to crush fruit or herbs in a glass or shaker for drinks like the Mojito.

Bottle opener

Choose a bottle opener with two attachments, one for metal-topped bottles and a corkscrew for wine bottles.

Pourers

A pourer is inserted into the top of a spirit bottle to enable the spirit to flow in a controlled manner.

Food processor

A food processor or blender is useful for making frozen cocktails and smoothies.

Equipment for garnishing

Exotic drinks may be prettified with a paper umbrella and several long drinks are served with straws or swizzle sticks.

TECHNIQUES

With just a few basic techniques, your bartending skills will be complete. Follow the instructions to hone your craft.

Blending

Frozen cocktails and smoothies are blended with ice in a blender until they are of a smooth consistency. Be careful not to add too much ice as this will dilute the cocktail. It's best to add a little at a time.

Muddling

A technique used to bring out the flavours of herbs and fruit using a blunt tool called a muddler.

1. *Add chosen herb(s) to a highball glass. Add some sugar syrup and some lime wedges.*
2. *Hold the glass firmly and use a muddler or pestle to twist and press down.*
3. *Continue for 30 seconds, top with crushed ice and add the remaining ingredients.*

Shaking

The best-known cocktail technique and probably the most common. Used to mix ingredients thoroughly and quickly, and to chill the drink before serving.

1. *Half-fill a cocktail shaker with ice cubes, or cracked or crushed ice.*
2. *If the recipe calls for a chilled glass add a few ice cubes and some cold water to the glass, swirl it around and discard.*
3. *Add the ingredients to the shaker and shake until a frost forms on the outside.*
4. *Strain the cocktail into the glass and serve.*

Double-straining

To prevent all traces of puréed fruit and ice fragments from entering the glass, use a shaker with a built-in strainer in conjunction with a hawthorne strainer. A fine strainer also works well.

Layering

Some spirits can be served layered on top of each other, causing lighter spirits to float on top of your cocktail.

1. *Pour the first ingredient into a glass, taking care that it does not touch the sides.*
2. *Position a bar spoon in the centre of the glass, rounded part down and facing you. Rest the spoon against the side of the glass as your pour the second ingredient down the spoon. It should float on top of the first liquid.*
3. *Repeat with the third ingredient, then carefully remove the spoon.*

Stirring

Used when the ingredients need to be mixed and chilled, but also maintain their clarity. This ensures there are no ice fragments or air bubbles throughout the drink. Some cocktails require the ingredients to be prepared in a mixing glass, then strained into the serving glass.

1. *Add the ingredients to a glass, in recipe order.*
2. *Use a bar spoon to stir the drink, lightly or vigorously, as described in the recipe.*
3. *Finish the drink with any decoration and serve.*

INDEX